Grade 3 · Unit 4

Observing Weather

FRONT COVER: (t)XING ZHOU/Moment/Getty Images, (b)mrdoomits/iStock/Getty Images;
BACK COVER: mrdoomits/iStock/Getty Images

Mheducation.com/prek-12

STEM McGraw-Hill is committed to providing instructional materials in Science, Technology, Engineering, and Mathematics (STEM) that give all students a solid foundation, one that prepares them for college and careers in the 21st century.

Send all inquiries to:
McGraw-Hill Education
8787 Orion Place
Columbus, OH 43240

ISBN: 978-0-07-699631-5
MHID: 0-07-699631-X

Printed in the United States of America.

4 5 6 7 8 9 10 11 LWI 26 25 24 23 22 21 20

Table of Contents
Unit 4: Observing Weather

Weather Impacts

Weather Impacts

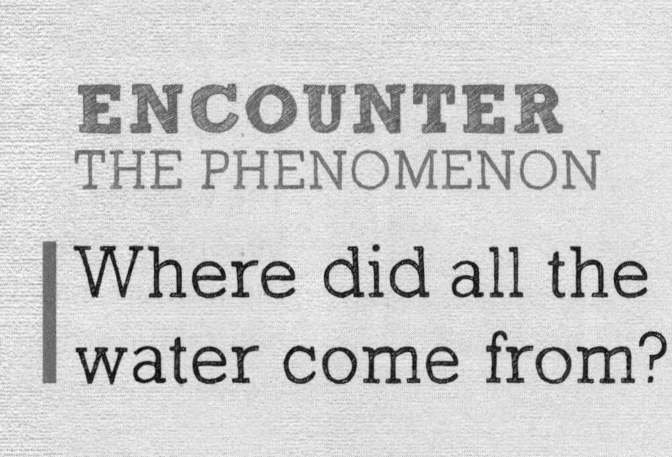

ENCOUNTER THE PHENOMENON

Where did all the water come from?

Outrageous Weather

GO ONLINE

Check out *Outrageous Weather* to see the phenomenon in action.

Talk About It

Look at the photo and explore the digital activity. What questions do you have about the phenomenon? Talk about your observations with a partner.

Did You Know?

Other planets in our solar system also get rain. It rains acid on Venus, but the acid evaporates before ever reaching the ground.

STEM Module Project
Science Challenge

Meteorologist for a Day

The weather center needs your help. You will become a meteorologist for a day. At the end of the module, create a weather report on one natural hazard. In your report, include how your viewers can prepare for the hazard.

Lesson 1
Weather Patterns

Lesson 2
Weather and Seasons

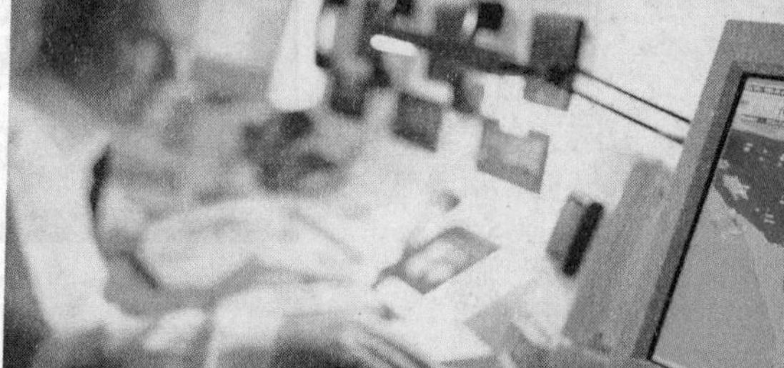

Meteorologists use many tools to collect data about the weather. They have the important job of warning people about severe weather.

Lesson 3
Natural Hazards and the Environment

Lesson 4
Prepare for Natural Hazards

What do you think you need to know to create a weather report?

HUGO
Meteorologist

STEM Module Project

Plan and Complete the Science Challenge Use what you learn throughout the module to plan a weather report.

LESSON 1 LAUNCH

Clouds in the Sky

When George woke up, there were dark clouds in the sky. On his way to school, it rained. When he walked home from school with his friends, the Sun was shining. There were white clouds in the sky. George and his friends had different ideas about the clouds.

This is what they said:

George: *I think the dark clouds are the only ones that contain water. When they get very dark, the water falls as rain.*

Kamila: *I think dark and white clouds contain water. Only the dark clouds release the water as rain.*

Jenna: *I don't think any clouds can hold water if the Sun is shining. White clouds do not contain water.*

Which friend do you agree with most? ______________________

Explain why you agree.

____

____

____

____

You will revisit the Page Keeley Science Probe later in the lesson.

LESSON 1

Weather Patterns

ENCOUNTER
THE PHENOMENON

How is it sunny in one place and cloudy in another?

GO ONLINE

Check out *Moving Clouds* to see the phenomenon in action.

Talk About It

Look at the picture and watch the video *Moving Clouds*. What kind of weather is the picture showing? What kind of weather will this place have next? Draw a picture to show your prediction.

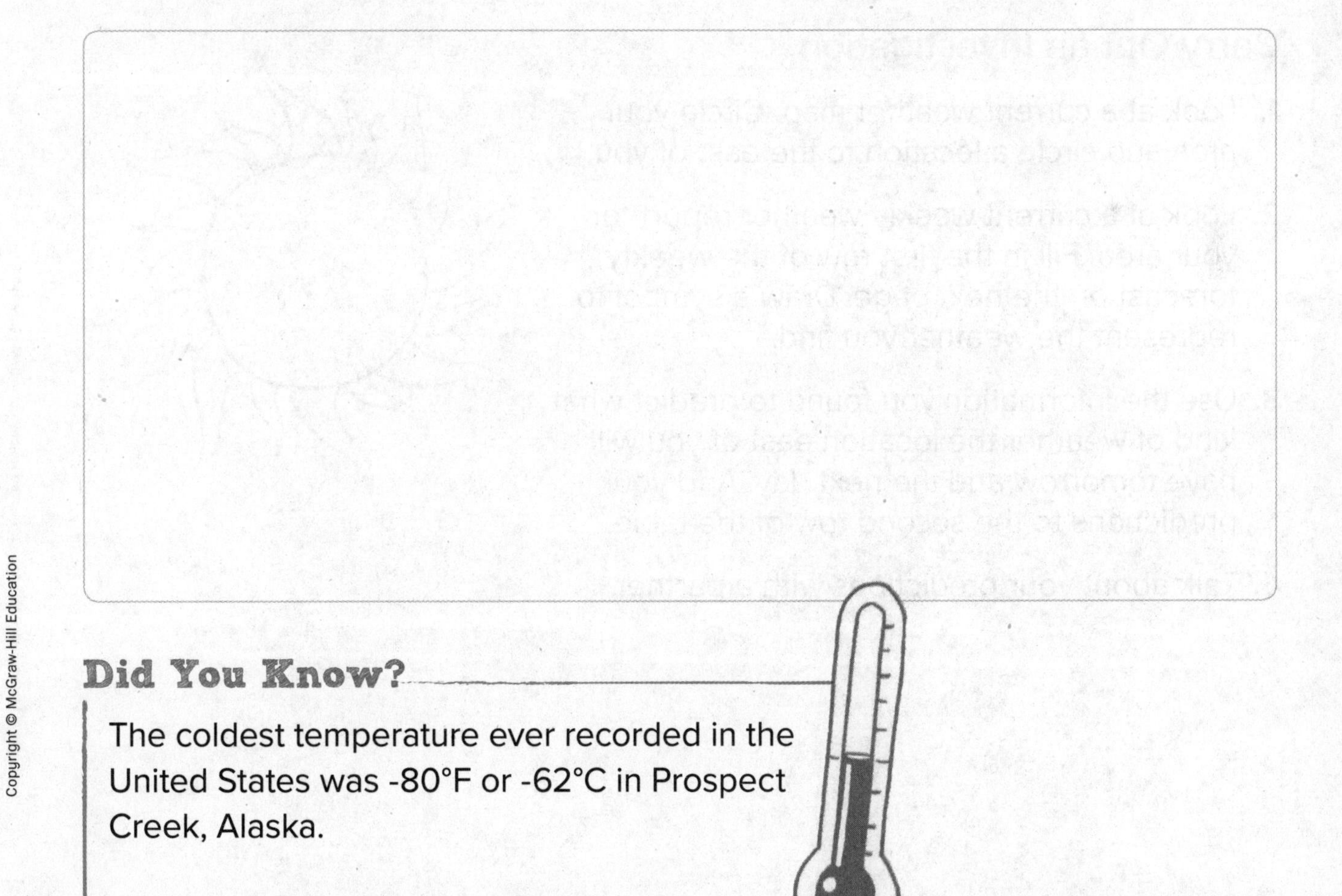

Did You Know?

The coldest temperature ever recorded in the United States was -80°F or -62°C in Prospect Creek, Alaska.

INQUIRY ACTIVITY

Data Analysis

Predict Weather

You looked at a picture of changing weather. Investigate to discover how weather can be predicted. Explain how this happens.

Materials

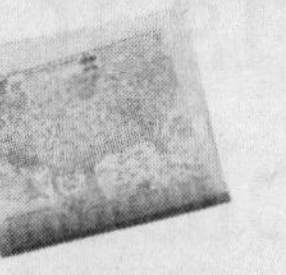

current weather map

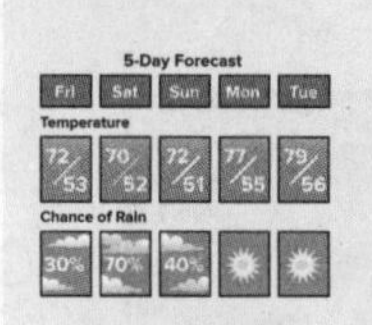

weekly weather report of your area

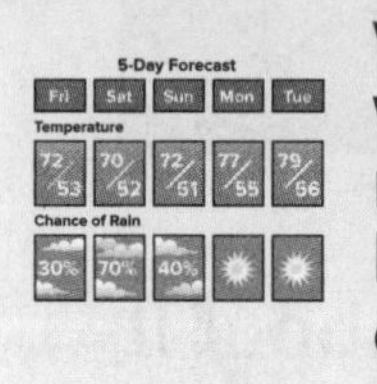

weekly weather report of location east of you

Make a Prediction What kind of weather will the area to the east of your location have tomorrow?

____

____

____

Carry Out an Investigation

1. Look at a current weather map. Circle your area and circle a location to the east of you.
2. Look at a current weekly weather report for your area. Fill in the first row of the weekly forecast on the next page. Draw a symbol to represent the weather you find.
3. Use the information you found to predict what kind of weather the location east of you will have tomorrow and the next day. Add your predictions to the second row of the table.
4. Talk about your predictions with a partner.

Weekly Forecast					
Area	Day:___	Day:___	Day:___	Day:___	Day:___
My Area:					
Area to the East:					

Communicate Information

6. What patterns did you notice between the weather report for your area and the one for the area east of you? How can you use this relationship to predict weather in the United States?

Talk About It

With a partner, discuss your predictions for the area east of you. Are they the same as the weather report? Why do you think this is?

Weather

VOCABULARY

Look for these words as you read:

atmosphere

precipitation

temperature

weather

Weather is what the air is like outside at a certain time and place. Even though you cannot see air you can see it move things, such as leaves of trees. Weather changes from day to day. It can also change from hour to hour.

The air that surrounds Earth is part of the atmosphere. The **atmosphere** is a blanket of gases and tiny bits of dust that surround Earth. The atmosphere has several layers. Weather occurs in the layer closest to Earth.

Temperature

When people describe weather, they describe the condition of the sky. Some terms used are sunny, cloudy, rain, and stormy. They also describe air and its **temperature**. Temperature is a measure of how hot or cold something is. A thermometer is a tool that measures temperature. The diagram on the next page shows how to read a thermometer. A higher temperature means it is getting warmer. When the temperature goes down, it gets colder.

When we describe weather, we talk about the clouds in the sky, the temperature of the air, and how the wind is blowing.

1. **MATH Connection** The height of the red bar, inside the thermometer, shows the temperature of the surrounding air. Look at the diagram. What is the temperature in degrees Celsius?

GO ONLINE
Watch the video *What is Weather?* to find out more about weather.

2. Place an arrow on the thermometer pointing to the temperature it is in your area today. Label the arrow.

3. What kind of activities can you do in this temperature? What should you wear when you go outside?

4. Water usually freezes at 0°C. What is the temperature at which water freezes in Fahrenheit?

5. **READING Connection** What makes up Earth's atmosphere?

Describing and Measuring Weather

Air temperature is one measurement of weather. Precipitation, wind, and air pressure also describe weather. When one of these factors changes, so does the weather.

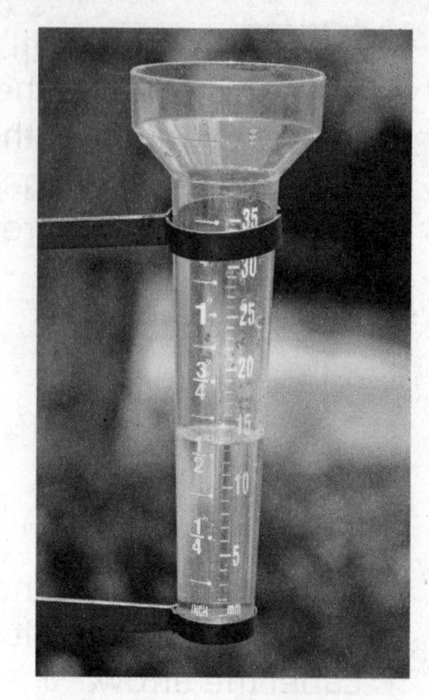

A rain gauge measure how much precipitation has fallen.

Precipitation **Precipitation** is water that falls to the ground from clouds. Liquid rain is the most common type of precipitation. Precipitation falls as liquid rain the air temperature is warmer than 0° C (32°F). A rain gauge is used to measure precipitation.

Sleet, snow and hail are frozen precipitation. Sleet forms when rain falls through a layer of freezing-cold air. Snow is made of ice crystals. Hail forms when rain freezes and is tossed about in a tall cloud.

Air Pressure Air pressure is the force of air pressing down on Earth's surface. Weather reports often describe air pressure. Scientists use a tool called a barometer to measure air pressure.

Wind Wind is moving air. It is caused by differences in air pressure. Scientists measure how fast the wind is blowing with an anemometer. Because winds in the United States usually blow from west to east, the weather will usually move from west to east as well.

What is the relationship between wind and air pressure?

The arrow of a weather vane points into the wind.

Predicting Weather

Weather balloons are used to gather data about weather.

Knowing the weather helps people stay safe. Airplane pilots study the weather to find out if it is safe to take off and land in another area. Predictions of hurricanes can give people time to find shelter.

Scientists use many tools to predict the weather. For example, a weather balloon is a tool that is launched into the air. It carries devices that collect data about the atmosphere. A satellite is a tool that scientists put into space. It travels around Earth and collects data over very large areas. A satellite can spot storms over the ocean. Scientists then use other information about the atmosphere to predict where the storm will move next.

Scientists use the data they collect and past information about an area to predict what the weather will be like in that area. They may look at what the weather is mostly like during a particular month or even a year.

Talk About It

What are some ways knowing weather predictions can help people plan their lives better?

Weather Map

A weather map lists the temperatures, precipitation, and other upcoming weather predictions over a large area. It may show the high temperatures for the day. It may show where it will be sunny.

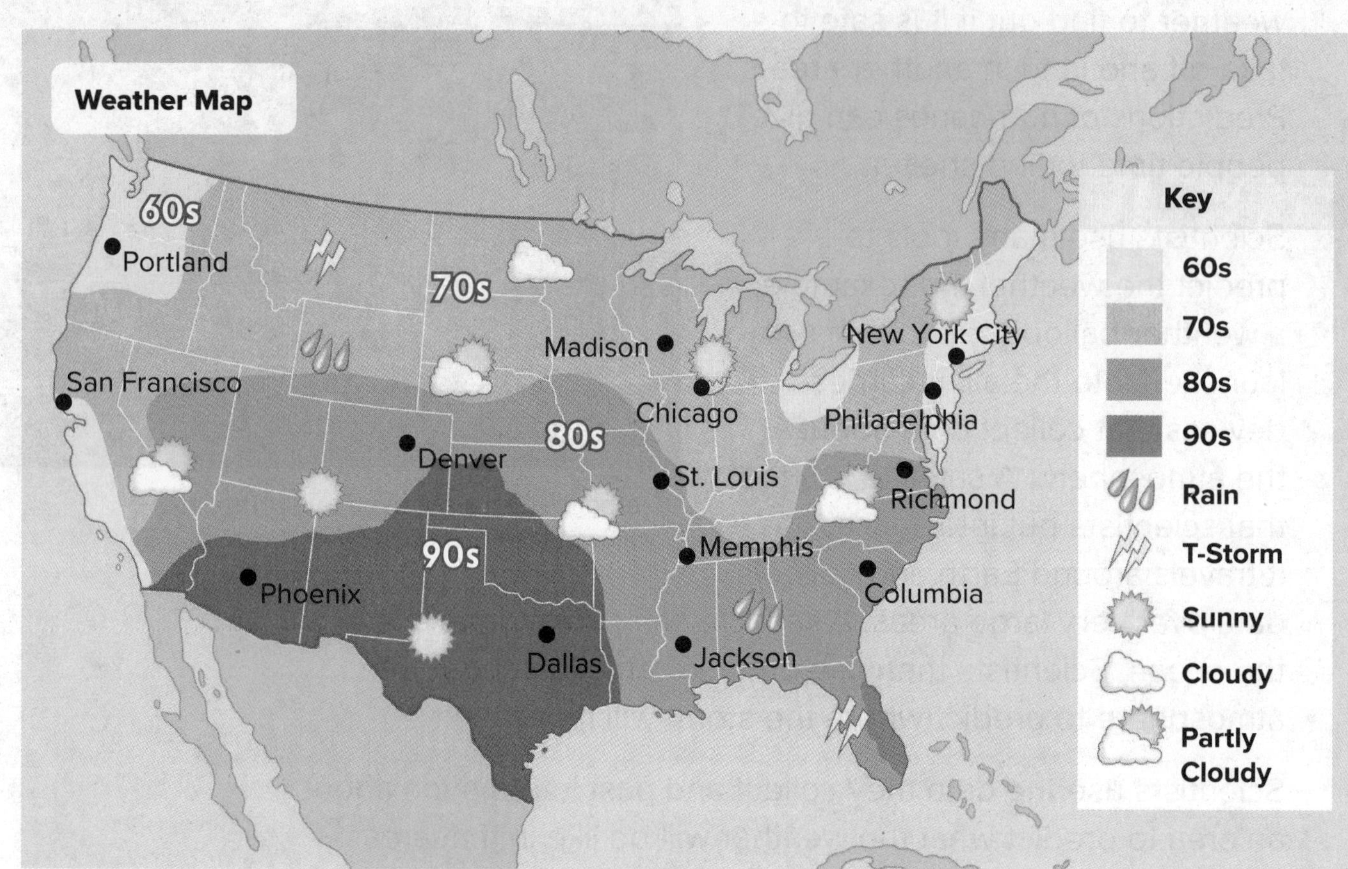

Read a Diagram Look carefully at the key. The colors show areas where different high temperatures will occur. The symbols show the kind of weather different areas will have.

Look at the weather map. Name two cities that will have high temperatures in the 90s.

Use the **data**. What will the **weather** be like in Chicago?

STEM Connection

What Does a Broadcast Meteorologist Do?

Broadcast Meteorologists report on the weather. They usually work for radio or television networks. Meteorologists study the effects of weather in their local communities and around the country. They read weather charts and information on past weather conditions to create weather reports. They report predictions for precipitation, temperatures, and wind. They are very good at communicating their data and findings. During weather broadcasts, they use maps and charts to explain weather patterns to people and make suggestions for how to prepare for the days ahead.

It's Your Turn

As a broadcast meteorologist, what information would you need to have in order to create a broadcast about your local weather? What would you show people to help them understand the forecast?

INQUIRY ACTIVITY

Research

Become a Meteorologist

Use weather maps and other information to write and present a weather forecast.

Materials

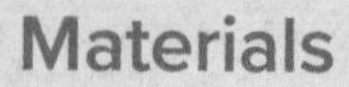

weather maps

charts and other weather information

State the Claim Which patterns will help in forecasting the weather?

Carry Out an Investigation

1. Study the weather maps, data charts, and other information.
2. **Record Data** On a separate sheet of paper, create a graph showing the information you have collected.

Communicate Information

3. Write your weather forecast.

GO ONLINE Explore *Patterns of Weather* to see how weather moves and changes.

4. What patterns did you notice in the weather?

Describe how **patterns** helped you make your **weather** forecast.

REVISIT Revisit the Page Keeley Science Probe on page 5.

EXPLAIN THE PHENOMENON

How is it sunny in one place and cloudy in another?

Summarize It

Explain how we can make predictions about weather in the United States.

____

____

____

____

____

____

____

REVISIT

Revisit the Page Keeley Science Probe on page 5. Has your thinking changed? If so, explain how it has changed.

Three-Dimensional Thinking

Study the weather map and look at the weekly forecast for the city of Dallas, Texas. Then answer questions 1–2 below.

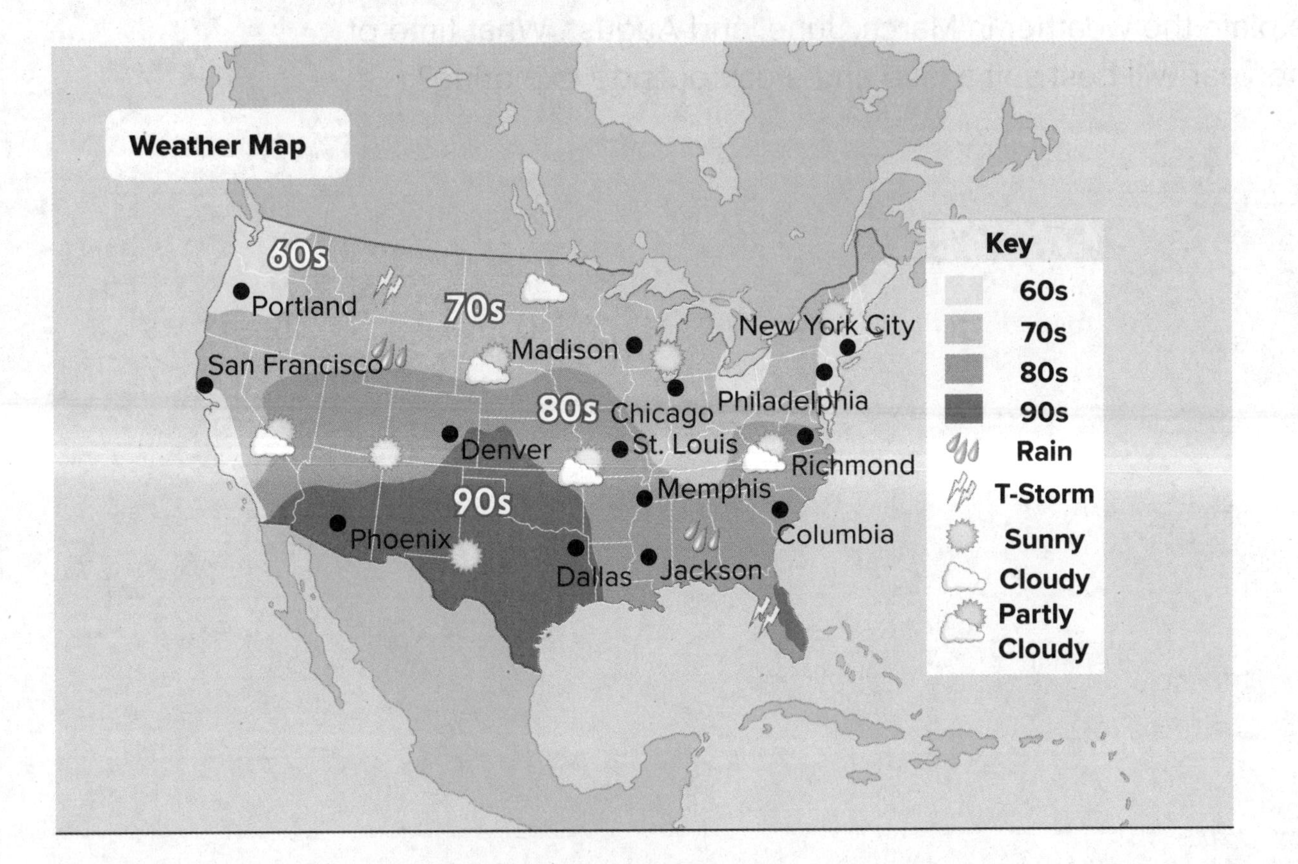

Weekly Forecast for Dallas, Texas:				
Monday	Tuesday	Wednesday	Thursday	Friday
Sunny	Sunny	Partly Cloudy	Partly Cloudy	Rain

1. The weather map predicts that the high temperature in Dallas will be in the ________.

   A. 60s　　B. 70s

   C. 80s　　D. 90s

2. What kind of weather conditions would you expect Dallas to have on Thursday?

Extend It

You have been asked to plan a class trip to a state park to study the local ecosystems. Weather plays a big role in deciding a day for the trip. Research the best time to head to the state park. Explain the weather in March, June, and August. What time of the year will best suit a hike and a day outside exploring?

KEEP PLANNING

STEM Module Project

Science Challenge

Now that you have learned patterns of weather, go to your Module Project to explain how the information will affect your weather report.

LESSON 2 LAUNCH

Weather Changes

Emily lives in San Diego, California. Her cousins live in Portland, Maine. When Emily went to visit her cousins in January, she noticed it was much colder than San Diego. Emily and her cousins had different ideas why it was much colder. This is what they said:

Emily: *I think temperature depends on location.*

Olivia: *I think temperature depends on where the Sun is.*

Petra: *I think temperature depends on location and where the Sun is.*

Tanya: *I think temperature depends on the time of day and location.*

Which friend do you agree with the most? ______________________

Explain why you agree.

____

____

____

You will revisit the Page Keeley Science Probe later in the lesson.

LESSON 2

Weather and Seasons

ENCOUNTER
THE PHENOMENON

Why does the tree change throughout the year?

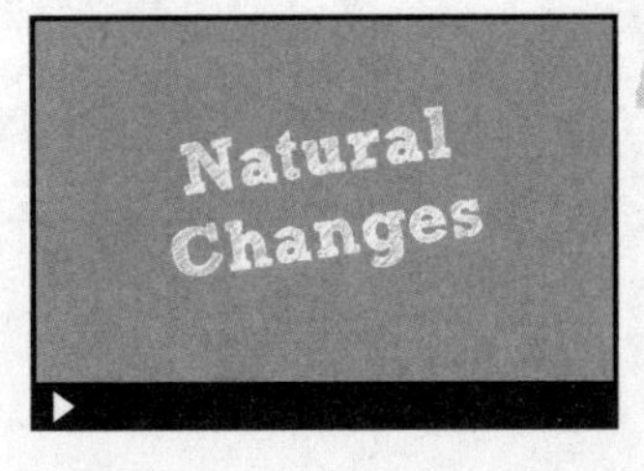

GO ONLINE

Check out *Natural Changes* to see the phenomenon in action.

Talk About It

Look at the picture and watch the video *Natural Changes*. Talk to a partner about the weather in the four photos. Record or illustrate your thoughts below.

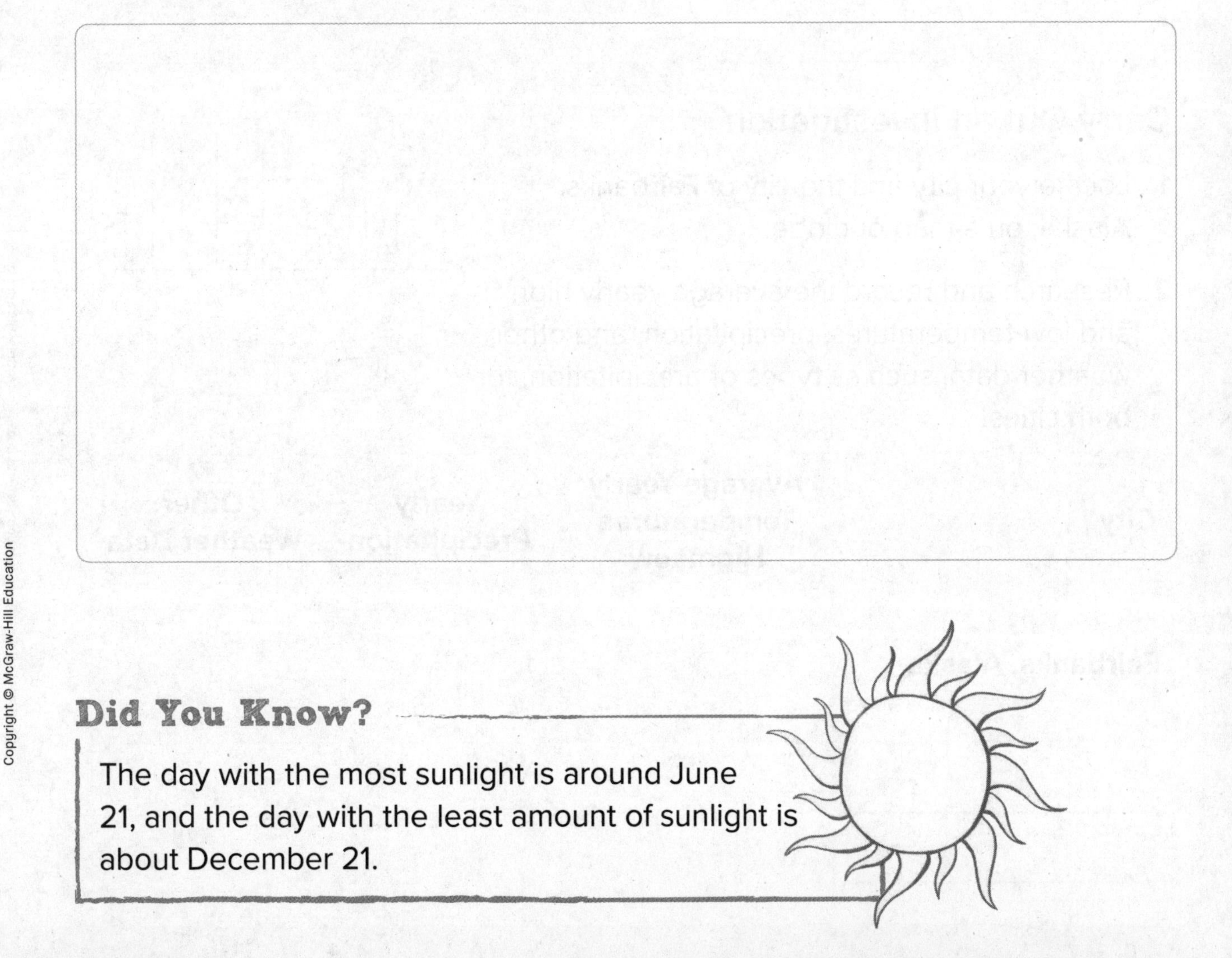

Did You Know?

The day with the most sunlight is around June 21, and the day with the least amount of sunlight is about December 21.

INQUIRY ACTIVITY

Research

Compare Weather Patterns

Materials

maps or globe

You learned that areas to the east of you will have similar weather to what you had before. You will investigate how weather in the north compares to the weather where you live.

State the Claim How do the temperatures and the precipitation in Fairbanks, Alaska, compare to the temperatures and precipitation in your city?

Carry Out an Investigation

1. Locate your city and the city of Fairbanks, Alaska, on a map or globe.
2. Research and record the average yearly high and low temperatures, precipitation, and other weather data, such as types of precipitation, for both cities.

City	Average Yearly Temperatures High/Low	Yearly Precipitation	Other Weather Data
Fairbanks, Alaska			

Communicate Information

3. How do the temperatures and precipitation amounts compare between the two cities?

4. Which city had the highest average yearly temperature? Which city had the lowest amount of yearly precipitation?

5. Did the results of your investigation support your claim? Explain.

Talk About It

What can you infer about weather patterns in northern areas? Do you think this is true for all northern areas? Explain.

VOCABULARY Look for these words as you read:

axis **climate** **season**

Climate

Weather changes all the time. It may be rainy one day and sunny the next. But the climate of the area stays the same. **Climate** is the pattern of weather in a certain place over a long period of time. A climate is described by its average temperature and precipitation. One area may have cool, dry summers. Another may have hot, humid summers.

Climates differ based on where an area is located on Earth. Not all areas have four separate seasons. Many areas have hot and cold temperatures and wet and dry periods throughout the year.

1. What is climate?

2. What is a city's climate based on?

Seasons

You saw the changing trees in the phenomenon. Trees change due to seasons. **Seasons** are times of the year with different weather patterns. Earth's four seasons are winter, spring, summer, and fall. The north and south halves of Earth have opposite seasons at any given time. The seasons are caused by the way Earth is tilted and moves around the Sun.

Winter is the coldest season. The Sun's path is lower in the sky. There are fewer hours of daylight. Temperatures can be cold in winter. Precipitation may fall as snow in some areas. Winter is too cold for some animals.

In spring, the Sun's path begins to rise higher. Temperatures are warm. There are more hours of daylight. Animals that were away during winter begin to return.

During the summer, there are more hours of sunlight during the day. The summer Sun is higher in the sky than at other times of the year. Temperatures are the warmest of the year.

In fall, there are fewer hours of daylight and temperatures are cooler. The Sun stays lower in the sky compared to summer.

Look at the photos of the seasons. Label each season.

GO ONLINE Check out the video *Winter and Summer Weather* to see the seasons in action.

Graphing Temperature

When you interpret data, you use the information that has been gathered to answer questions or to solve problems. It is often easier to interpret data when it is shown in a table or a graph.

1. **MATH Connection** Use the data table to make a line graph.

Average Monthly Air Temperature in Sacramento, California (°C)											
Jan.	**Feb.**	**Mar.**	**Apr.**	**May**	**June**	**July**	**Aug.**	**Sept.**	**Oct.**	**Nov.**	**Dec.**
12	16	19	23	26	30	35	32	29	25	17	12

2. Analyze the data in the table and in your line graph. Which months are coolest? Which months are warmest?

Earth and Climate

Earth is the shape of a sphere, or ball. Earth also has an imaginary line called an **axis** through its center. Earth is constantly moving around this axis, like a spinning top. However, Earth's axis is tilted slightly. Earth's axis also points to the same place in the sky all year long. This consistent slant and direction of Earth's axis as it orbits the Sun causes different seasons. This affects climates around the world.

Because Earth is shaped like a ball, incoming solar rays strike Earth at different angles depending upon where you live. Closer to the equator, the mid-day Sun is high overhead. These places receive more energy per square foot and generally have higher temperatures and warmer climates. Farther north or south, the mid-day Sun is lower in the sky. Here less energy is received per square foot and mid-day temperatures are generally cooler resulting in colder climates.

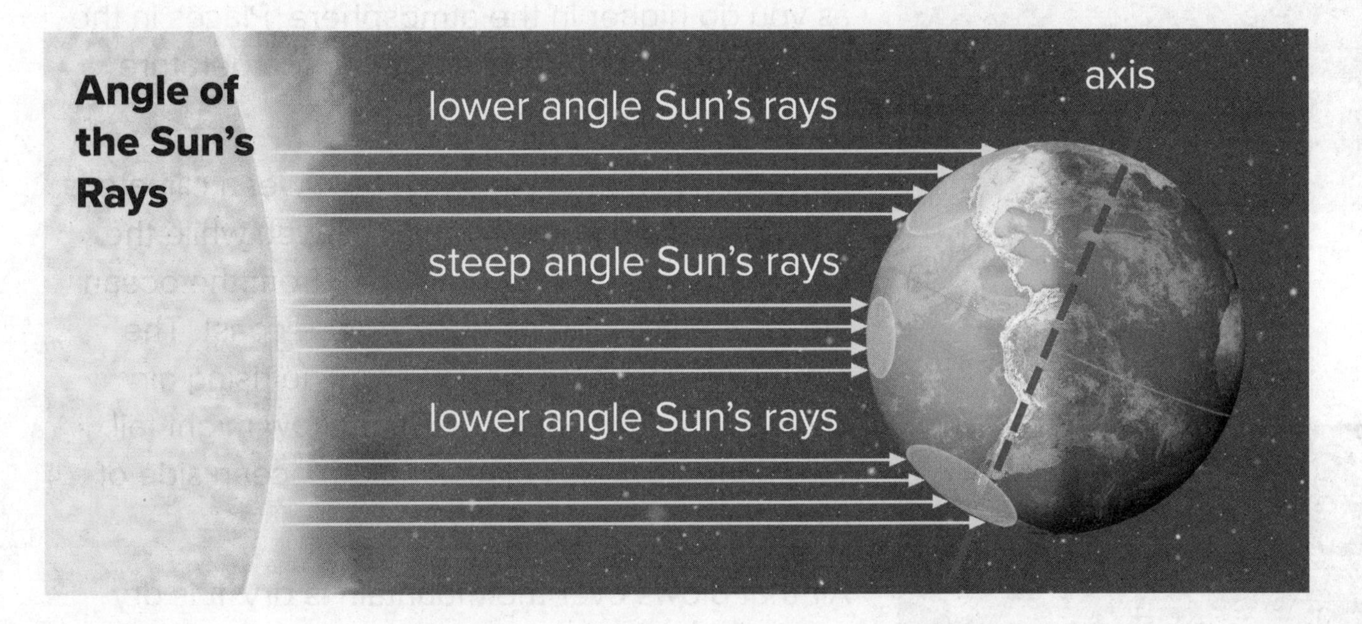

Revisit the Page Keeley Science Probe on page 21.

Factors that Affect Climate

Seattle, Washington, is near the ocean. It has milder temperatures and more rain than places farther inland.

Water Being near an ocean or other large body of water affects climate. Water absorbs and gives off energy more slowly than land. In summer, ocean water is cooler than nearby land. This tends to keep the air above the land cooler. In winter, ocean water is warmer than nearby land. The air above the land near the ocean is warmer than land farther inland.

Large lakes also affect climate. Air blowing across lakes can pick up moisture. The moisture can fall as rain or snow on land areas near the lake.

Breckenridge, Colorado, is high in the Colorado Rockies. It has cool temperatures.

Height How high in the atmosphere a place is affects its climate. Air temperatures get colder as you go higher in the atmosphere. Places in the mountains tend to have colder air temperature and climates than lower areas.

Mountains Mountains affect how wet a climate is. One side of a mountain might be wet, while the other side might be dry. Moist air from the ocean moves toward mountains along the coast. The mountains force the air upward. The rising air cools and forms clouds. Rain or snow might fall. This pattern causes places on the ocean side of mountains to have a wet climate.

Air cools and loses moisture as it moves up and over a mountain.

Air that blows over the mountain is dry. It is dry because the air has lost its moisture on the ocean side. Dry air blows down this side of the mountain. It is common to see deserts on the dry side of a mountain. The mountains block moist air from reaching inland.

Circle the text that explains why many places near the ocean have mild climates.

Cut out the Notebook Foldables tabs given to you by your teacher. Glue the anchor tabs as shown below. Use what you have learned to make notes about the different climates.

Compare Data

GO ONLINE Explore *Compare Data* to see weather data in action.

Investigate finding patterns in climate data by conducting the simulation. Complete the data chart after you have explored.

1. Choose two of the cities that are close to latitude 30° south. Complete the data chart.

City	Average High Temperature	Average Low Temperature	Warmest Month

What **patterns** do you see in the **data** for the two cities?

2. Choose two different cities. Complete the data chart.

City	Average High Temperature	Average Low Temperature	Warmest Month

What **patterns** do you see in the **data** for these two cities?

3. Compare your data charts with a classmate's charts. What patterns do you see?

STEM Connection

What Does A Climatologist Do?

Climatologists are scientists who study the climate. They study weather patterns over long periods of time. Instead of looking at a ten-day weather forecast, they look at climate changes over years—or even decades! This data is collected by weather satellites.

Have you heard of global change? Climatologists research this problem. They are interested in our health and the health of plants and animals, too. They are also very interested in oceans and how they affect the climate.

Most climatologists like working with other people. This is a good thing because climatologists usually work in teams. They have big problems to solve.

It's Your Turn

How is a climatologist's job different from a meteorologist's job? How do climatologists help people prepare for—or avoid—natural disasters?

INQUIRY ACTIVITY

Hands On

Land and Temperature Change

Explore how the color of a land's surface affects how much heat it absorbs from the Sun.

Make a Prediction If ______________________________

____

______________ then ______________________________

______________________________________ because

____.

Materials

____

____

____

____

____

____

____

Carry Out an Investigation

BE CAREFUL Wear safety goggles at all times.

1. Write a list of the materials you will use.

2. Plan your procedure.

____

____

____

____

____

____

____

____

3. **Record Data** Create a table to show the data you collect.

4. **Analyze Data** In your table, circle the type of land that warmed up the most in the given time.

Communicate Information

5. How could the color of the soil affect the local climate?

6. Did the results of your investigation support your prediction? Explain.

LESSON 2

Review

EXPLAIN THE PHENOMENON

Why does the tree change throughout the year?

Summarize It

Explain why the seasons change throughout the year.

REVISIT Revisit the Page Keeley Science Probe on page 21. Has your thinking changed? If so, explain how it has changed.

Three-Dimensional Thinking

1. The state of Nevada has a dry climate because it is on the dry side of the Sierra Nevada mountain range. Which state would you expect to have a similar climate to Nevada?

A. Utah

B. Florida

C. Washington

D. Wisconsin

2. Explain why the state you chose in Question 1 will have a similar climate to Nevada.

3. Explain two of the four seasons.

Extend It

You have entered a poster contest. The theme is Seasons. Draw the patterns that occur when the seasons change. Present your poster to the class.

KEEP PLANNING

STEM Module Project
Science Challenge

Now that you have learned about weather and the seasons, go to your Module Project to explain how the information will affect your weather report.

LESSON 3 LAUNCH

Habitat Hazards

Four friends were talking about changes to habitats that can harm the organisms that live there. They each had a different idea about what causes the harmful changes. This is what they said:

Sofia: *I think harmful changes to habitats are caused by hazards created by humans.*

Dean: *I think harmful changes to habitats are caused by natural hazards.*

Hoda: *I think harmful changes to habitats are caused by both hazards created by humans and natural hazards.*

Alfie: *I think harmful changes to habitats are caused by hazards created by humans. Natural hazards cause changes to habitats, but they are not harmful.*

Who do you agree with the most? ______________________

Explain why you agree.

____

____

____

You will revisit the Page Keeley Science Probe later in the lesson.

LESSON 3

Natural Hazards and the Environment

ENCOUNTER
THE PHENOMENON

What affected the plant growth in this area?

GO ONLINE

Check out *Wildfire* to see the phenomenon in action.

Talk About It

Look at the photo and watch the video *Wildfire*. What questions do you have about the phenomenon? Talk about them with a partner. Record or illustrate your thoughts below.

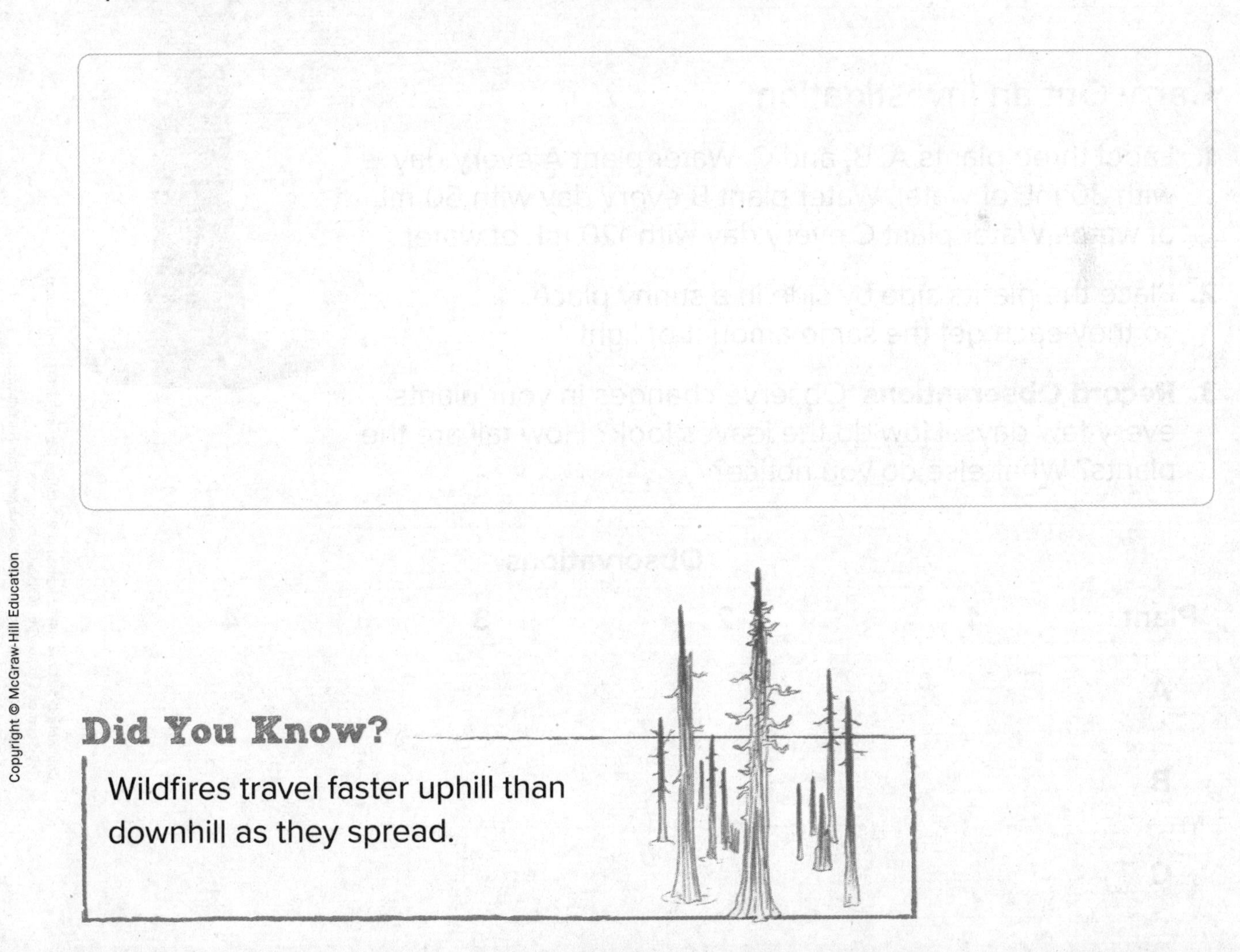

Did You Know?

Wildfires travel faster uphill than downhill as they spread.

INQUIRY ACTIVITY

Hands On

Flooding Plants

Materials

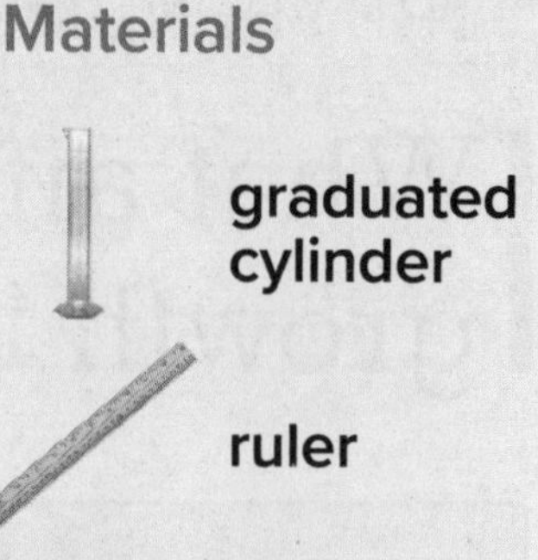

graduated cylinder

ruler

3 pansy plants of the same size

Wildfires are one kind of natural disaster. Floods are another. When an area receives more rain than usual, a flood can occur. A flood can affect the living things in an area. What happens to plants when there is a flood?

Make a Prediction What will happen to a plant that gets too much water?

____

____

____

Carry Out an Investigation

1. Label three plants A, B, and C. Water plant A every day with 30 mL of water. Water plant B every day with 60 mL of water. Water plant C every day with 120 mL of water.
2. Place the plants side by side in a sunny place so they each get the same amount of light.
3. **Record Observations** Observe changes in your plants every few days. How do the leaves look? How tall are the plants? What else do you notice?

	Observations			
Plant	**1**	**2**	**3**	**4**
A				
B				
C				

Communicate Information

4. Draw a picture of the three plants after two weeks.

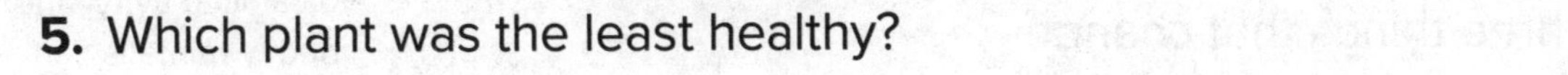

5. Which plant was the least healthy?

6. What evidence could you use to argue about which plant is the least healthy?

7. Construct an Explanation How can a flood affect plants?

Talk About It

Did your findings support your prediction? Explain why or why not to a partner.

VOCABULARY Look for these words as you read:

natural hazard

Environmental Changes

Some changes on Earth happen quickly. Fast changes on Earth can cause great damage to the environment. A **natural hazard** is a natural event, such as a flood, earthquake, or hurricane, that causes great damage. Natural disasters are usually weather related. Wildfires are also a type of disaster. They are typically caused by a lightning strike. Some natural disasters are more common in some areas than others. Unfortunately, there is not a way to avoid natural disasters. Knowing what disasters may occur in your area can help you prepare and protect your family and friends.

1. What are three things that change environments?

2. **ENVIRONMENTAL Connection** What might happen if there is a long period of no rain?

Earthquakes, volcanic activity, and heavy rainfalls can cause soil and rocks to move down a slope.

GO ONLINE Watch the video *Environmental Changes* to see what happens to environments after a natural disaster.

Earthquake

An earthquake is the shaking of Earth's surface. Earthquakes are caused by movement in Earth's crust. The crust is made of huge slabs of rock. The slabs can slowly move past each other and can press against each other. They can pull apart too. These movements happen along large cracks called faults. When the crust suddenly moves along these faults, the area above the fault can shake.

GO ONLINE Check out *Measuring Disasters* to see how to measure disasters.

An earthquake can cause more than just a rumble. It causes vibrations at the place where the two plates hit each other. The vibrations move out in all directions at once. The areas closest to the center of the quake have the strongest vibrations. The areas farther away have weaker vibrations.

WRITING Connection Obtain and combine information from several resources to explain how different technologies can reduce the damage caused by earthquakes. Use the Summarize graphic organizer to help write your response. Share your summary with the class.

Floods

Water can slowly change the land, or it can change the land quickly. Sometimes, more rain falls than can soak into the ground. Water quickly fills rivers and streams. If a river cannot hold all the water, the water flows over its banks, causing a flood. A flood is water that flows over land that is normally dry. Rapid melting of snow can also cause flooding. Floods can be very strong and dangerous. The heavy flow of water can wash away large objects, such as cars.

Floods can change the shape of the land by eroding soil quickly. Even after the flood is over, the land may be reshaped. A river's course may be changed. Buildings, bridges, roads, and crops may be washed away. Sediment carried by the floodwater may have been deposited on land.

Landslides

A landslide is the rapid movement of rocks and soil down a hill or mountain. Heavy rocks, trees, and land are pulled down the hill by gravity. Landslides can happen quickly and cause damage to property. Landslides may occur when soil loosens because of heavy rain or melting snow. Some landslides may occur without warning. Others can happen after earthquakes shake loose large areas of rock or soil on hillsides.

Landslides can happen on land or the ocean floor. Landslides on hills and mountains can destroy homes and roads. They can bury large areas. Underwater landslides can form giant waves.

INQUIRY ACTIVITY

Hands On

Landslide

You will create a model of a landslide. Landslides occur mostly in hilly areas. They are usually triggered by an earthquake or flash flood.

Make a Prediction What would happen if a large portion of land moved over an area of homes?

____

____

Carry Out an Investigation

BE CAREFUL Wear safety goggles at all times.

1. Place one book under the top of the plastic paint tray.
2. Use 2 cups of dry sand. Pour the sand along the top of the paint tray to 2 cm deep.
3. Place a small cube every 2 cm along the sand to represent houses.
4. Slowly pour water onto the sand. Observe how the flooding affects the sand. Tap the bottom of the paint tray to model an earthquake.

Communicate Information

5. Describe what happened to the sand.

____

____

Talk About It

What precautions can people take to reduce the loss and damage caused by landslides?

Materials

- safety goggles
- book
- plastic paint tray
- sand
- ruler
- gram cubes

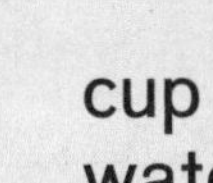

- cup of water
- cup

Inspect

Read the passage *Growing Up in Tornado Alley.* Underline the text that explains what a tornado is.

Find Evidence

Reread How does Joshua know that a tornado is near? Highlight the text evidence.

Notes

Growing Up in Tornado Alley

"A tornado is scary," explains ten-year-old Joshua Amerman from Borger, Texas. Joshua lives in an area called Tornado Alley. This Great Plains region has the perfect weather conditions to form tornadoes.

A tornado is a violent, rotating column of air that stretches from the bottom of a thunderstorm to the ground. Its powerful, circular winds can blow over 300 miles per hour.

"Once, when I was younger," Joshua remembers, "I was playing with my brother in the backyard. A storm blew in and the clouds were an eerie, dark green color. The wind began to whip up dirt and leaves. Suddenly, my mom called us inside." A large tornado was headed toward their home. "My brother and I lay down in our bathtub, and my mom covered us with the mattress off my bed." I was so frightened, and I worried about my pets."

The tornado sirens blared. The wind roared like a train. Outside, the sky was as dark as night. Then, a calm settled in and all was quiet. As quickly as it formed, the tornado had disappeared. Outside Joshua's house, however, the tornado had made a mess. Joshua's favorite tree was torn from the earth. Their swimming pool was crumpled. Joshua's family was shaken, but thankfully unharmed.

This was not the only time Joshua has hidden from a tornado. And it probably won't be his last. "If you live in Texas, you learn about tornadoes," Joshua declares. Meteorologist Robert Slattery of the National Weather Service in Amarillo, Texas, agrees. A tornado can cause great destruction. He suggests that everyone be educated in tornado safety.

Revisit the Page Keeley Science Probe on page 39.

Make Connections

Talk About It

Look again at the photos in this passage. What are the similarities and differences between a tornado and an earthquake? Share your thoughts with your partner.

Notes

INQUIRY ACTIVITY

Research

Natural Hazards

GO ONLINE Watch the video *Natural Hazards* to learn more about different characteristics of weather events.

Watch *Natural Hazards* on the various weather events. When do we call something a natural disaster?

State the Claim How can you determine whether or not a recent event was a natural disaster?

____

____

Carry Out an Investigation

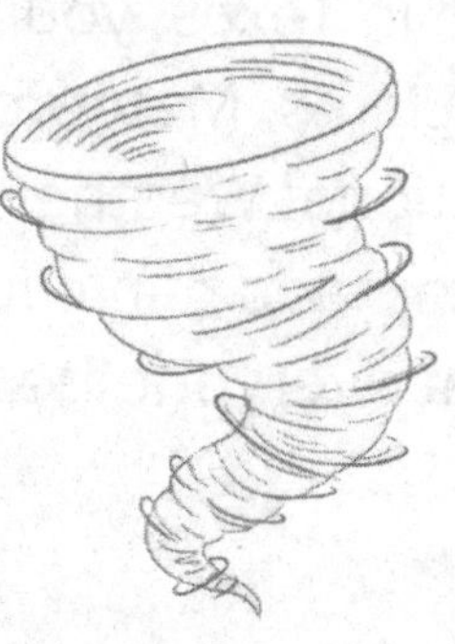

1. Choose a recent weather event.
2. **Research** how the event happened and what caused it.
3. Circle the type of event you chose below.

Communicate Information

What **evidence** can you use to argue that it was a **natural hazard**?

____

____

____

____

STEM Connection

What Does a Hydrologist Do?

Hydrologists study groundwater as it moves through rocks and soil underground. This is an important job because people need groundwater to drink. Farmers also need groundwater to grow their crops.

Hydrologists make sure the water is not contaminated. Water pollution is a big problem. Humans accidentally pollute the water supply by using too many pesticides or with landfills or septic tanks. Hydrologists help solve water problems. They work closely with hydrologists who study surface water, including rivers, lakes, and oceans.

It's Your Turn

As a hydrologists who studies groundwater, how might you work with a hydrologist who studies surface water? How do you think hydrologists help solve water problems?

LESSON 3

Review

EXPLAIN THE PHENOMENON

What affected the plant growth in this area?

Summarize It

Explain how natural hazards affect the environment.

REVISIT Revisit the Page Keeley Science Probe on page 39. Has your thinking changed? If so, explain how it has changed.

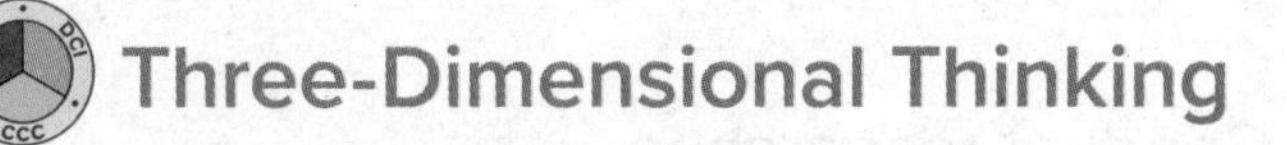

Three-Dimensional Thinking

1. Circle all the pictures that show evidence of a natural hazard.

A. 

B. 

C. 

D. 

2. Which natural disaster could also be caused by an earthquake?

A. tornado

B. forest fire

C. landslide

D. flashflood

3. Joel builds a hill of sand in his backyard. He places some sugar cubes on the hill. He pours a bucket of water down the hill and observes what happens to his model. He most likely does this to see ___________.

A. how a flashflood affects the land

B. how a forest fire affects the land

C. how a tornado affects the land

D. how a rain shower affects the land

Extend It

You are an expert on natural hazards. Conduct research to explain how climate affects the occurrence of a natural hazard event in the continental United States. Include information on how to stay safe during this disaster. Prepare a multimedia presentation to present your findings to the class

KEEP PLANNING

STEM Module Project
Science Challenge

Now that you have learned about natural hazards, go to your Module Project to explain how the information will affect your weather report.

LESSON 4 LAUNCH

Natural Hazards

Natural hazards impact how humans and other organisms live.
Put an X in any of the boxes that best describe natural hazards.

Natural hazards can result from natural processes.	Humans cause most natural hazards.	Humans can stop most natural hazards.
Humans can reduce the impact of natural hazards.	Natural hazards are helpful because they are natural.	Scientists can predict some natural hazards.
Scientists can prevent most natural hazards.	Scientists study natural hazards.	Natural hazards depend on weather conditions.

Explain your thinking. Describe your ideas about natural hazards.

____

____

____

____

____

You will revisit the Page Keeley Science Probe later in the lesson.

LESSON 4

Prepare for Natural Hazards

ENCOUNTER
THE PHENOMENON

How can I stay safe in severe weather?

GO ONLINE

Check out *Storm* to see the phenomenon in action.

Talk About It

Look at the photo and watch the video of the *Storm*. What kind of weather is the photo showing? Talk to your partner about your observations. Record or illustrate your thoughts below.

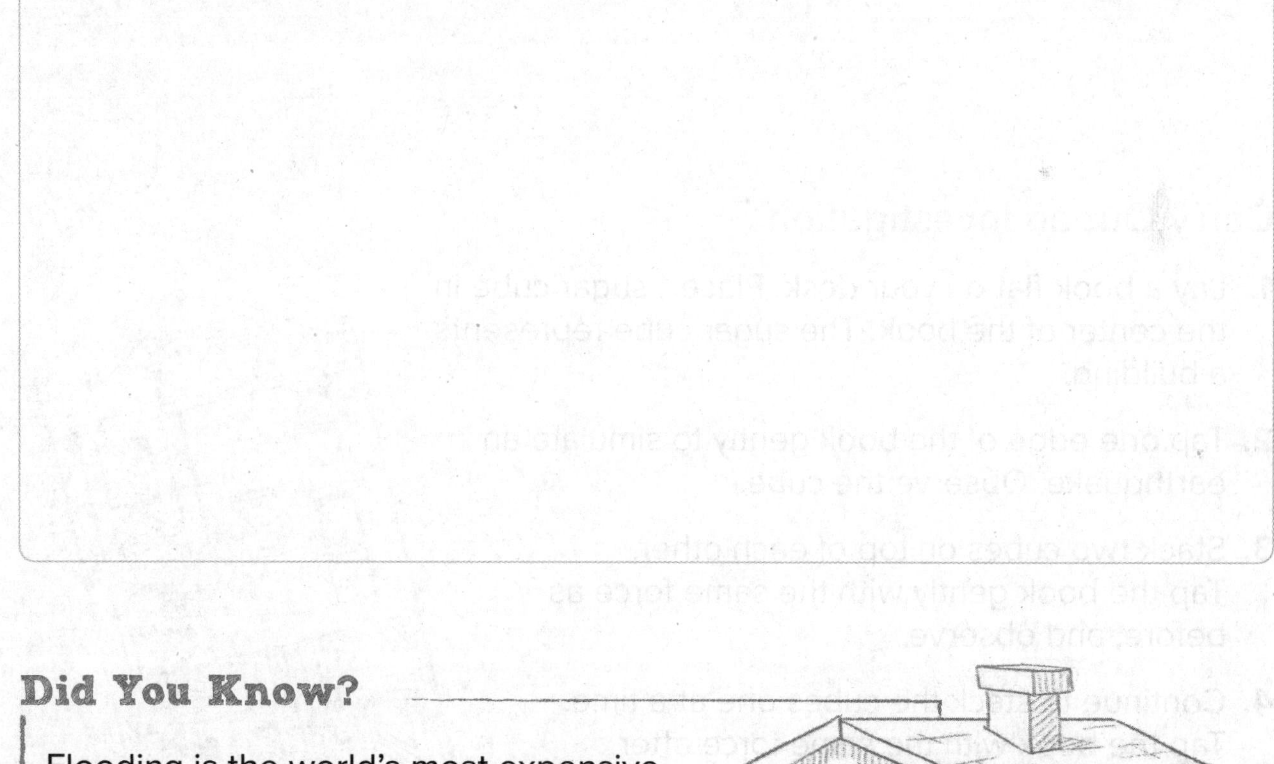

Did You Know?

Flooding is the world's most expensive type of natural disaster. Floods cause a lot of damage to buildings and cities.

INQUIRY ACTIVITY

Hands On

Build Sugar Structures

In the video, you observed how rain affects buildings. Model and observe how earthquakes affect buildings.

Make a Prediction How many sugar cubes can be stacked without falling over when they are tapped?

Carry Out an Investigation

1. Lay a book flat on your desk. Place 1 sugar cube in the center of the book. The sugar cube represents a building.
2. Tap one edge of the book gently to simulate an earthquake. Observe the cube.
3. Stack two cubes on top of each other. Tap the book gently with the same force as before, and observe.
4. Continue to stack the cubes one at a time. Tap the book with the same force after you add each cube. See how many cubes you can stack before the stack falls over.

Sugar Cubes Stacked	
Number of Cubes	**Observations**

Communicate Information

5. How did your model building stand up to the shaking?

6. Did your findings support your prediction? Explain.

INQUIRY ACTIVITY

7. Use the evidence that you collected to tell how your model might show how an earthquake affects a building.

Talk About It

How would you modify the criteria and constraints of the investigation to improve your model?

MAKE YOUR CLAIM

How does the design of a building reduce the impact of natural hazards?

Make your claim. Use your investigation.

CLAIM

_______________ are more likely to withstand a natural hazard.

Cite evidence from the activity.

EVIDENCE

I found that _______________.

Discuss your reasoning as a class. Tell about your discussion.

REASONING

The evidence supports the claim because _______________.

You will revisit your claim later in the lesson.

Scientists Study Natural Hazards

VOCABULARY

Look for these words as you read:

floodwall

levee

lightning rod

How can scientists solve problems and help people? One way is by studying natural hazards. Studying hurricanes helps scientists learn how strong buildings should be built to keep people safe in high winds. There are many ways to study natural hazards. In wildfire labs, scientists study how fires grow and move. They might design new tools that put out fires faster and easier.

Studying earthquakes helps us find out where to build the strongest buildings or bridges. Scientists can make stronger, more flexible materials.

Scientists study the effects of earthquakes on structures.

Floods and hurricanes are also important natural hazards to study. How can a strong, high wall keep water from reaching a community? There are two kinds of walls. One is a floodwall. **Floodwalls** are walls built to reduce or prevent flooding in an area. The other is a levee. A **levee** is a wall built along the side of rivers and other bodies of water to prevent them from overflowing. The work of scientists who study natural hazards combined with the work of engineers who design and build structures helps to keep us safe.

Scientists work with engineers to build floodwalls.

Create a Venn diagram and describe the similarities and differences between a floodwall and a levee.

GO ONLINE Watch the video *Humans and Natural Disasters* to learn more about how to reduce the effects of natural disasters.

Building Structures

GO ONLINE Explore *Building Structures* to see different structures that are made to protect.

Engineers use the information from scientists to design and build structures. Explore the simulation to see the different parts of a building that keep us safe during an earthquake.

What are two parts of a building that engineers use to help **keep the structure and people safe during an earthquake?**

Lightning Rods

Another structure that helps prevent damage from natural hazards is a lightning rod. A **lightning rod** is a metal bar that safely directs lightning into the ground. During thunderstorms, these help keep houses and buildings safe. They were first thought of by Benjamin Franklin in the 1750s. He experimented with electricity and lightning using a kite and key. Later, he used what he discovered to find a way to keep houses safe.

INQUIRY ACTIVITY

Hands On

Sandbags and Floods

Learn how to prevent damage from a flood. People use sandbags to reduce the effect of water on the land.

Make a Prediction How will the sandbags change the effect of the water on the land?

If sandbags are used to reduce the amount of water that flows over the land, then

____

____

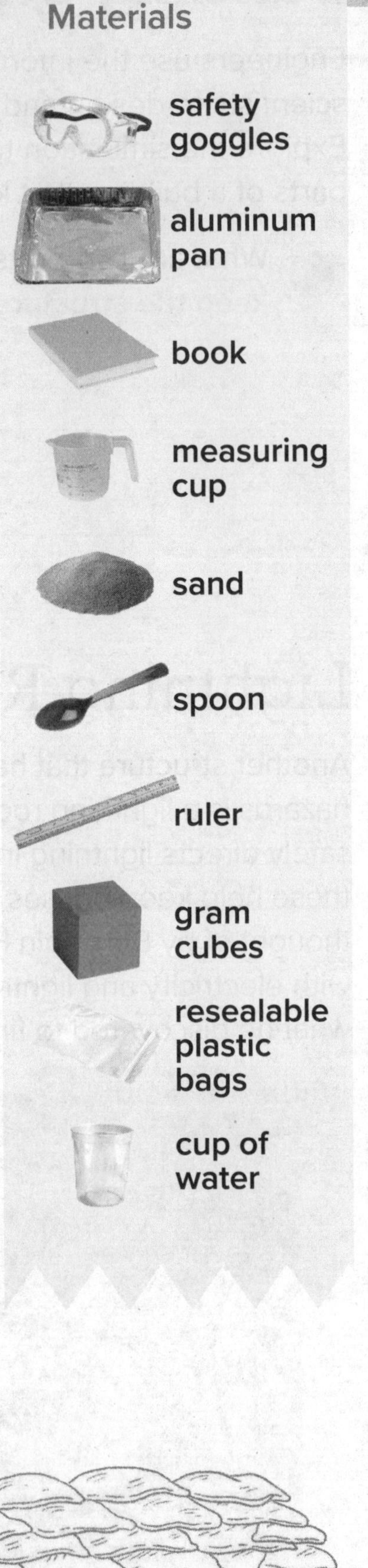

Carry Out an Investigation

BE CAREFUL Wear safety goggles to protect your eyes from the sand.

1. Place the book under one side of the pan.
2. Place dry sand along the top half of the pan, about 2 cm deep. Use several cups of sand.
3. Place a small cube every 2 cm along the sand to represent houses.
4. Make your sandbags by filling each bag with 4 to 5 spoonfuls of sand. Seal the bag to hold the sand in.
5. Build a wall out of the sandbags. Position your wall so that it blocks the water from flowing over the part of your pan that has the houses.
6. Slowly pour water out of the cup onto the sand. Observe how the flooding affects the sand.

Communicate Information

7. Did your findings support your prediction? Explain.

8. **Make an Argument** Use your observations from the activity to tell how well the wall of sandbags worked to reduce the effects of a landslide.

COLLECT EVIDENCE

Add evidence to your claim on page 61 about how the design of a building can reduce the impact of a natural hazard.

Inspect

Read the passage *Preparing for Natural Hazards*. Underline text that tells what supplies people can gather in the event of a disaster.

Find Evidence

Reread Find and highlight the text that means almost the same as *hazard*.

Notes

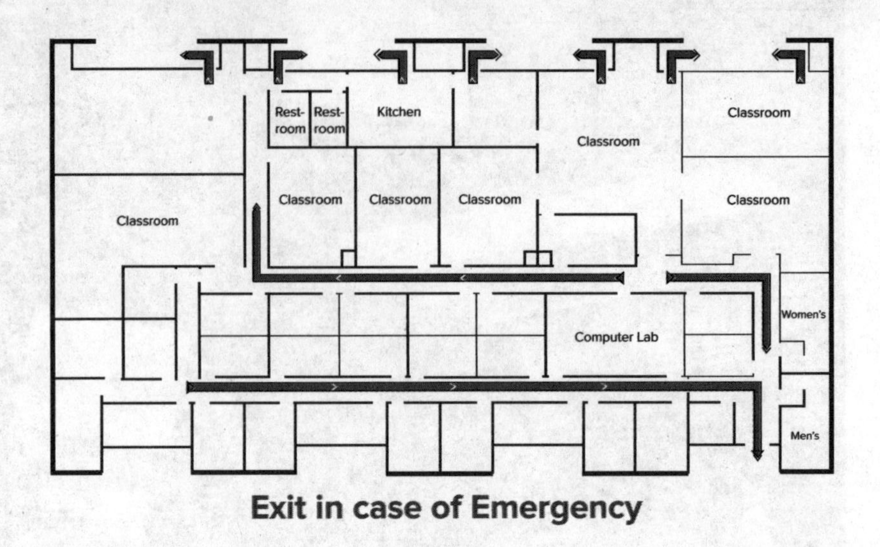

Exit in case of Emergency

Classrooms might have an exit map to show students how to exit the school quickly and safely in case of an emergency.

Preparing for Natural Hazards

What does it mean to prepare for a disaster? To prepare means to find out information and be ready. If you are prepared for natural hazards, you can protect yourself and your belongings.

The first step to being prepared is research. What type of natural hazards do you need to be aware of in your area? Some areas of the country may experience severe flooding while other areas may have wildfires sweep through their neighborhoods. Preparing for a blizzard is very different than preparing for a landslide.

Another step in preparing for natural hazards is to have a plan. Knowing how to react before, during, and after an event can save lives. Making a plan with your family will help you contact one another and reconnect if separated. Establish a family meeting place that's familiar and easy to find.

News reporter warns viewers of storms that are coming.

Television stations are prepared to alert people if a natural disaster is coming. They will explain what is expected. There are alerts that are sent over the radio that can update those that are traveling. People with mobile phones can receive alerts about weather and other disasters as well.

People can plan ahead for emergencies. Many people have a radio with batteries. They have stored water and emergency supplies. It is also smart to memorize important phone numbers. If they know a storm is coming, families might get extra food so that they don't have to travel during the storm.

Schools are prepared as well. Many classrooms have maps posted in their classroom. Students and teachers practice how to exit the building safely or learn what areas of the building are safest, depending on the natural hazard.

REVISIT PAGE KEELEY SCIENCE PROBES Revisit the Page Keeley Science Probe on page 55.

Make Connections

Talk About It

What are some natural hazards that could occur in your area? Research and discuss with your partner.

WRITING Connection You are an engineer hired by a city. The mayor has asked you to compare three design solutions to prevent and reduce water damage caused by flooding. The city, which is located near a river, has many residents and not many open spaces. Research and evaluate three designs and choose the best design solution for the mayor's city. Use evidence to support your claim.

STEM Connection

What Does A Civil Engineer Do?

Civil Engineers design and build large structures, such as buildings, bridges, and dams. They know a lot about construction and how to build structures that can survive natural disasters. They also know a lot about architecture.

You might be surprised to know that some civil engineers spend a lot of time studying and thinking about water. They design plans for people to get over water or through water—or to stop water from hurting people. They are also responsible for building systems to clean water and get rid of water pollution.

It's Your Turn

Think like a civil engineer. Complete the next activity, and build a weatherproof structure.

INQUIRY ACTIVITY

Engineering

Build Weatherproof Structures

As a civil engineer, you will build and perform tests on a hurricane-proof model building. You will communicate your results and make an argument.

Define a Problem What are some things that affect the stability of a building during a hurricane?

____

____

____

____

Materials

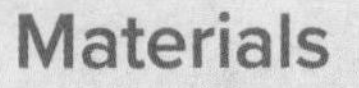

- safety goggles
- gelatin set in a plastic container

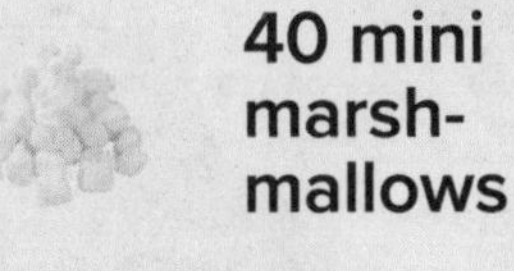

- 40 mini marsh-mallows
- 40 toothpicks

- tray
- fan

- watering can full of water
- modeling clay
- craft sticks

Carry Out an Investigation

BE CAREFUL Use caution when handling the toothpicks, as well as using the fan and water. Wear safety goggles at all times.

1. On a separate sheet of paper, draw and label a design for a sturdy building that has more than one level. It will need to withstand wind and rain, and it must be built out of up to 40 marshmallows and 40 toothpicks.
2. Carry out your plan and build your design. Construct your model building on top of the gelatin.
3. Place the container of gelatin with the model building in a large plastic tub. Your teacher will turn the fan on high with the air directed at the structure. Gently pour water over the structure for 30 seconds.

4. **Record Data** Record your observations below.

5. Make modifications to your model building. You may choose to use the other materials.

6. **Test Your Solution** Empty the plastic tub and place your new model building inside. Use the fan and watering can to apply wind and water to the building for 30 seconds.

7. **Record Data** Record your observations below.

Communicate Information

8. **Make an Argument** On a separate sheet of paper, use evidence that you collected in the activity to tell how well your improved model building met the requirements that you identified and solved the problem of withstanding wind and water.

9. How could your design be used to help humans reduce the impact of a storm surge?

10. Read the Investigator article, *Hurricane*. What other methods can be used to protect a building from hurricanes?

LESSON 4

Review

EXPLAIN THE PHENOMENON

How can I stay safe in severe weather?

Summarize It

Explain how people can prepare for natural hazards.

REVISIT Revisit the Page Keeley Science Probe on page 55. Has your thinking changed? If so, explain how it has changed.

Three-Dimensional Thinking

1. What structure will help a building withstand an earthquake best?

A. A tall wooden building

B. A one-story brick building

C. A tall brick building

D. A short wooden building

2. What is a natural hazard that is possible where you live? What can you do to be protected during this hazard?

3. Explain how you can protect structures in an area from flooding.

Extend It

You have been asked to prepare an emergency bucket for your classroom. Conduct research to discover what types of items you will need in case of an emergency. List your items and the costs of these items below.

OPEN INQUIRY

What questions do you still have?

Plan and carry out an investigation to answer one of the questions.

KEEP PLANNING

STEM Module Project
Science Challenge

Now that you have learned how to prepare for natural hazards, go to your Module Project to explain how the information will affect your weather report.

PERFORMANCE EXPECTATION

STEM Module Project
Science Challenge

Meteorologist for a Day

You are a meteorologist at a local news station. You have been asked to give a special report about natural hazards. Your goal is to present information about one hazardous weather condition and ways that people can prepare for the event.

Planning after Lesson 1

Apply what you have learned about weather patterns to your project planning.

How does knowing about patterns in weather help in your planning?

Record information to help you plan your weather report after each lesson.

STEM Module Project

Science Challenge

Planning after Lesson 2

Apply what you have learned about weather and the seasons to your project planning.

How does understanding climate help understand hazardous weather?

Planning after Lesson 3

Apply what you have learned about natural hazards and the environment to your project planning.

How does a natural hazard affect the environment?

Planning after Lesson 4

Apply what you have learned about how you prepare for natural hazards to your project planning.

How does preparing for a natural hazard keep you safe?

Research

Research your natural hazards and how to prepare for them using resources provided by your teacher or by finding books at your local library.

STEM Module Project
Science Challenge

Meteorologist for a Day

Look back at the planning you did after each lesson. Use that information to complete your final module project.

Carry Out an Investigation

1. Use your project planning to prepare your weather report.
2. Write out how you plan to present your weather report.
3. Determine if you need any materials. List the materials in the space provided.
4. You should use a variety of media or visual components, such as graphs, charts, images, video, or audio.

Materials

Sketch Your Project

In the space below, write or draw your report.

STEM Module Project
Science Challenge

Communicate Your Results

Share the plan for your project and your results with another group. Compare how you plan to prepare for the natural hazard. Communicate your findings below.

MODULE WRAP-UP

Using what you learned in this module, explain the outrageous weather.

Have your ideas changed? Explain.

Science Glossary

A

adaptation a structure or behavior that helps an organism survive in its environment

atmosphere a blanket of gases and tiny bits of dust that surround Earth

attract to pull toward

axis an imaginary line through Earth from the North Pole to the South Pole

B

balanced forces forces that cancel each other out when acting together on an object

birth the beginning or origin of a plant or animal

C

camouflage an adaptation that allows an organism to blend into its environment

climate the pattern of weather at a certain place over a long period of time

competition the struggle among organisms for water, food, or other resources

D

direction the path on which something is moving

distance how far one object or place is from another

E

ecosystem the living and nonliving things that interact in an environment

electrical charge the property of matter that causes electricity

environmental trait a trait that is affected by the environment

extinction the death of all of one type of living thing

F

floodwall a wall built to reduce or prevent flooding in an area

force a push or pull

fossil the trace of remains of living thing that died long ago

friction a force between two moving objects that slows them down

G

germinate to begin to grow from a seed to a young plant

group a number of living things having some natural relationship

H

hibernation to rest or go into a deep sleep through the cold winter

I

inherited trait a trait that can be passed from parents to offspring

instinct a way of acting that an animal does not have to learn

invasive species an organism that is introduced into a new ecosystem

L

learned trait a new skill gained over time

levee a wall built along the sides of rivers and other bodies of water to prevent them from overflowing

life cycle how a certain kind of organism grows and reproduces

lightning rod a metal bar that safely directs lightning into the ground

M

magnet an object that can attract objects made of iron, cobalt, steel, and nickel

magnetic field the area around a magnet where its force can attract or repel

magnetism the ability of an object to push or pull on another object that has the magnetic property

metamorphosis the process in which an animal changes shape

migrate to move from one place to another

mimicry an adaptation in which one kind of organism looks like another kind in color and shape

motion a change in an object's position

N

natural hazard a natural event such as a flood, earthquake, or hurricane that causes great damage

P

pole one of two ends of a magnet where the magnetic force is strongest

pollination the transfer of pollen from the male parts of one flower to the female parts of another flower

population all the members of a group of one type of organism in the same place

position the location of an object

precipitation water that falls to the ground from the atmosphere

R

repel to push away

reproduce to make more of their own kind

resource a material or object that a living thing uses to survive

S

season one of the four parts of the year with different weather patterns

static electricity the build up of an electrical charge on a material

survive to stay alive

speed a measure of how fast or slow an object moves

T

temperature a measure of how hot or cold something is

trait a feature of a living thing

U

unbalanced forces forces that do not cancel each other out and that cause an object to change its motion

V

variation an inherited trait that makes an individual different from other members of the same family

W

weather what the air is like at a certain time and place

Index

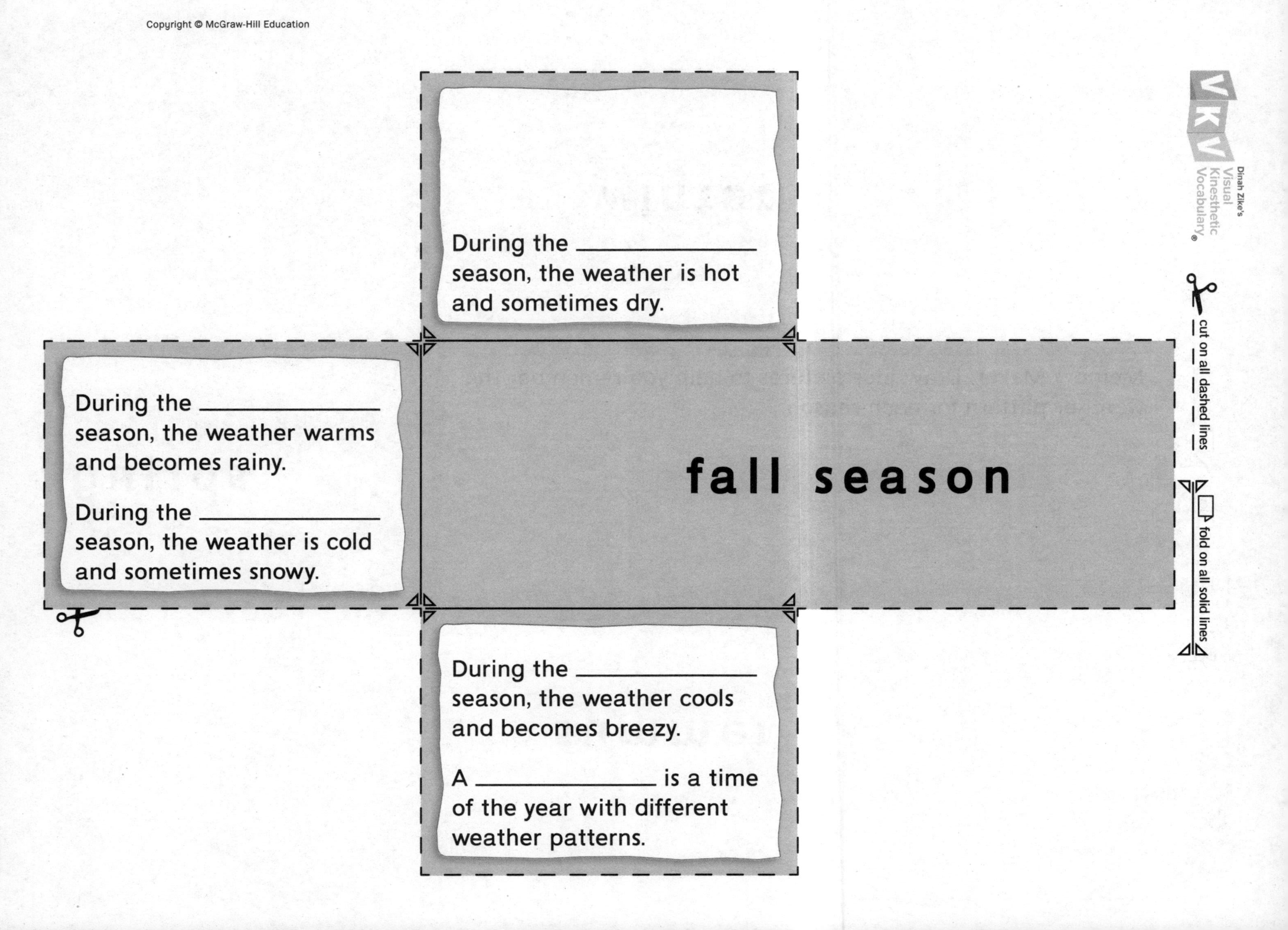
During the ______ season, the weather is hot and sometimes dry.
During the ______ season, the weather warms and becomes rainy.
During the ______ season, the weather is cold and sometimes snowy.
fall season
During the ______ season, the weather cools and becomes breezy.
A ______ is a time of the year with different weather patterns.
Dinah Zike's
Visual
Kinesthetic
Vocabulary®
VKV
cut on all dashed lines
fold on all solid lines

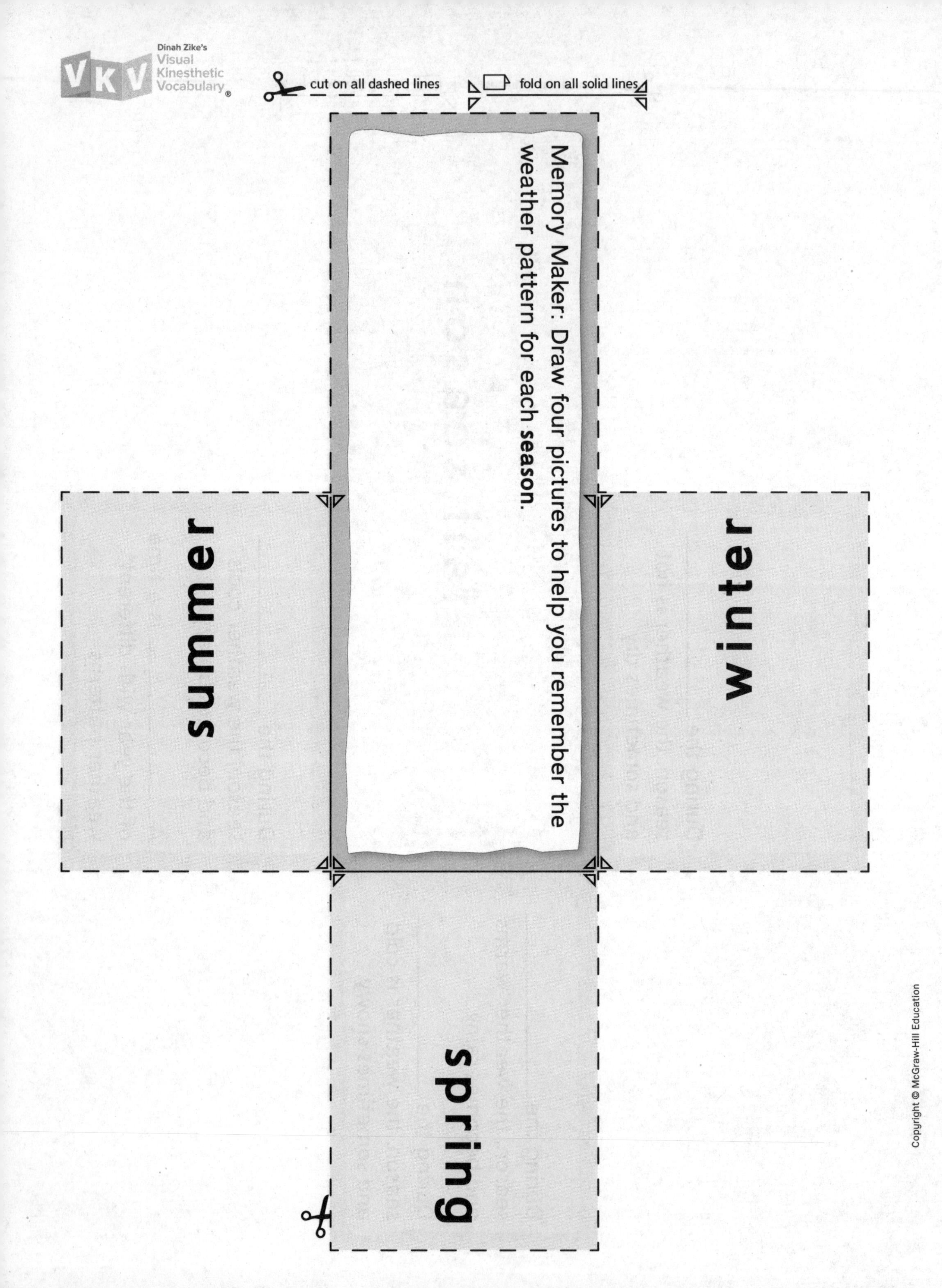
Dinah Zike's Visual Kinesthetic Vocabulary
VKV
cut on all dashed lines
fold on all solid lines
Memory Maker: Draw four pictures to help you remember the weather pattern for each **season**.
summer
winter
spring